NUMBERING & M
IN THE CLASSIC

NUMBERING & MEASURING
IN THE
CLASSICAL WORLD

An Introductory Handbook
(Revised 2nd Edition)
William F. Richardson

BRISTOL
PHOENIX
PRESS

First published in 1985 by St Leonards Publications, Auckland, New Zealand
Reprinted in 1992 by Bristol Classical Press

This revised edition published in 2004 by
Bristol Phoenix Press
an imprint of University of Exeter Press
Reed Hall, Streatham Drive
Exeter, Devon, EX4 4QR
UK

www.exeterpress.co.uk

Reprinted 2005
Printed digitally since 2008

A catalogue record for this book is available
from the British Library.

ISBN 978 1 904675 18 1

Printed and bound by CPI Group (UK) Ltd, Croydon, CR0 4YY

Author's Note

Whoever sets out to write on Greek and Roman weights and measures must produce either a very short book or a very long one. This is the very short one: it gives the basic facts as simply as possible and merely hints at the complexities that would form the bulk of the other.

I am grateful to John H. Betts of the Bristol Phoenix Press for the opportunity to produce a corrected and augmented edition. I wish him every success in his new endeavour.

Readers should note that all modern measures in this book are given in metric form; hence the terms 'foot', 'pound' and so on always refer to the ancient measures with these names.

William F. Richardson
Department of Classics and Ancient History
University of Auckland
Auckland, New Zealand

TABLE OF CONTENTS

CHAPTER ONE

THE CARDINAL NUMBERS

A. GREEK

1.1   The Greek cardinal numbers from one to nine hundred are shown in Table 1.1 (for the numerical symbols representing 1-900 see Chapter Four).

	units	tens	hundreds
1.	εἷς	δέκα	ἑκατόν
2.	δύο	εἴκοσι	διακόσιοι
3.	τρεῖς	τριάκοντα	τριακόσιοι
4.	τέσσαρες	τεσσαράκοντα	τετρακόσιοι
5.	πέντε	πεντήκοντα	πεντακόσιοι
6.	ἕξ	ἑξήκοντα	ἑξακόσιοι
7.	ἑπτά	ἑβδομήκοντα	ἑπτακόσιοι
8.	ὀκτώ	ὀγδοήκοντα	ὀκτακόσιοι
9.	ἐννέα	ἐνενήκοντα	ἐνακόσιοι

Table 1.1: The Greek cardinal numbers 1-900

Those that end in -οι are declined like μεγάλοι -αι -α. Εἷς is declined εἷς ἕνα ἑνός ἑνί in the masculine gender, μία μίαν μιᾶς μιᾷ in the feminine, and ἕν ἑνός ἑνί in the neuter. Δύο has a form δυοῖν for the genitive and dative cases. Τρεῖς has a neuter form τρία and in all genders has τριῶν and τρισί for genitive and dative respectively. Τέσσαρες in the masculine and feminine is declined τέσσαρες τέσσαρας τεσσάρων τέσσαρσι and in the neuter τέσσαρα τεσσάρων τέσσαρσι. The form τέτταρες is Attic. All the other numbers are indeclinable.

1

1.2   The intervening numbers are expressed by accumulating those in Table 1.1. Only ἕνδεκα 'eleven' and δυώδεκα 'twelve' are written as one word; beyond these the words are in ascending order linked by καί 'and', or in descending order with no καί. So Xenophon writes παρασάγγαι πέντε καὶ τριάκοντα καὶ πεντακόσιοι 'five hundred and thirty-five parasangs' (*Anabasis* II 2 6) and σταθμοὶ ἑκατὸν εἴκοσι δύο 'one hundred and twenty-two days' march' (*Anabasis* V 5 4).

1.3   To express one thousand and above Greek had the adjectives χίλιοι 'one thousand' and μύριοι 'ten thousand'. Compounds of each using the numeral adverbs (see 2.4) are found; they produce the sequence in Table 1.2.

1000	χίλιοι	10000	μύριοι
2000	δισχίλιοι	20000	δισμύριοι
3000	τρισχίλιοι	30000	τρισμύριοι
4000	τετρακισχίλιοι	40000	τετρακισμύριοι
5000	πεντακισχίλιοι	50000	πεντακισμύριοι
6000	ἑξακισχίλιοι	60000	ἑξακισμύριοι
7000	ἑπτακισχίλιοι	70000	ἑπτακισμύριοι
8000	ὀκτακισχίλιοι	80000	ὀκτακισμύριοι
9000	ἐνακισχίλιοι	90000	ἐνακισμύριοι

Table 1.2: The Greek cardinal numbers 1000-90000

1.4   Greek possessed a series of collective nouns ending in -άς, such as τριάς τριάδος '(a group of) three'. The powers of ten in this series (namely μονάς δεκάς ἑκατοντάς χιλιάς μυριάς) were frequently used for very large numbers: so the Greek for one hundred thousand is δέκα μυριάδες, and for one million ἑκατὸν μυριάδες (literally 'one hundred ten thousands'). This system appears frequently in the text of Herodotus for very large numbers: at VII 186 2 we find '5,283,220 men' expressed as 'five hundred and twenty-eight ten-thousands (μυριάδες) plus three thousands (χιλιάδες) plus two hundreds (ἑκατοντάδες) plus two tens (δεκάδες) of men'. The phrase μύριαι μυριάδες (literally 'ten thousand ten thousands' or 'ten thousand times ten thousand') is used in the New Testament and elsewhere for an indefinite but very large number.

1.5   The Greek mathematicians said that all the numbers from one to one hundred million (ten to the power of eight) were οἱ πρῶτοι ἀριθμοί 'the first numbers'. The number 'one hundred million' was then regarded as 'unity in the second numbers (οἱ δεύτεροι ἀριθμοί)'; the second numbers ran (in modern terminology) from 10^8 to 10^{16}.

1.6   In his treatise Ψαμμίτης (in Latin, *Arenarius*) Archimedes uses this as the basis of a system for naming astronomically large numbers. Under this system 200,200,002 would be expressed as 'two units in the second numbers (δύο μονάδες τῶν δευτέρων ἀριθμῶν) plus twenty myriads and two units in the first numbers (εἴκοσι μυριάδες καὶ δύο μονάδες τῶν πρώτων ἀριθμῶν'). This system continues up to οἱ μυριακισμυριοστοὶ ἀριθμοί 'the ten thousand times ten thousandth numbers', which end at (but do not include) a figure expressed in Arabic notation as 1 followed by eight hundred million noughts. The lower end of this system is set out in Table 1.3.

	οἱ πρῶτοι ἀριθμοί	οἱ δεύτεροι ἀριθμοί	οἱ τρίτοι ἀριθμοί
μονάς unity	10^0	10^8	10^{16}
δεκάς ten	10^1	10^9	10^{17}
ἑκατοντάς hundred	10^2	10^{10}	10^{18}
χιλιάς thousand	10^3	10^{11}	10^{19}
μυριάς ten thousand	10^4	10^{12}	10^{20}
δέκα μυριάδες	10^5	10^{13}	10^{21}
ἑκατὸν μυριάδες	10^6	10^{14}	10^{22}
χίλιαι μυριάδες	10^7	10^{15}	10^{23}
μύριαι μυριάδες	10^8	10^{16}	10^{24}

Table 1.3: Archimedes' system for astronomically large numbers

1.7   At this point in his system Archimedes adds a technical term, referring to the whole of this set of numbers as ἡ πρώτη περίοδος 'the first period', so that the figure mentioned at the end of 1.6 becomes μονὰς τῶν πρώτων ἀριθμῶν τῆς δευτέρας περιόδου 'unity in the first numbers in the second period'. The periods then go up to the ten thousand times ten thousandth period, by which time the numbers express a quantity larger than the number of grains of sand in the world.

B. LATIN

1.8   Table 1.4 shows the Latin cardinal numbers from one to nine hundred. The ones that end in –i (but not *uiginti*) are declined like *magni magnae magna*. *Tres* has *tria, trium* and *tribus*. *Unus* is declined like *magnus magna magnum*, including, curiously, the plural (used in phrases such as *una castra* 'one camp'). The forms of *duo* are set out in Table 1.5. All the other words (including *uiginti*) are indeclinable.

	units	tens	hundreds
1	unus	decem	centum
2	duo	uiginti	ducenti
3	tres	triginta	trecenti
4	quattuor	quadraginta	quadringenti
5	quinque	quinquaginta	quingenti
6	sex	sexaginta	sescenti
7	septem	septuaginta	septingenti
8	octo	octoginta	octingenti
9	nouem	nonaginta	nongenti

Table 1.4: The Latin cardinal numbers 1-900

1.9   For numbers such as twenty-four Latin has the same two patterns as English: *uiginti quattuor* 'twenty-four' and *quattuor et uiginti* 'four-and-twenty'. The numbers from eleven to seventeen are of the second type but without *et*: *undecim duodecim* and so on; for *duodeuiginti* 'eighteen' and *undeuiginti* 'nineteen' see 2.11. For compound numbers above 100 the same two patterns apply.

	masculine	feminine	neuter
nominative	duo	duae	duo
accusative	duos	duas	duo
genitive	duorum	duarum	duorum
dative	duobus	duabus	duobus
ablative	duobus	duabus	duobus

Table 1.5: The forms of *duo*.

1.10   For one thousand the Latin word is *mille*. This word, with its plural *milia*, is in origin a neuter noun like *cubile*, governing a partitive genitive as in *mille annorum* 'a thousand of years' (Plautus: *Miles Gloriosus* 1079) and *milia passuum octo* 'eight thousands of paces' (that is, eight miles: Caesar: *Gallic War* I 21 1). Aulus Gellius, in an essay on the subject (*Noctes Atticae* I 16), quotes an ablative singular form *milli* and also points out that *mille*, being grammatically singular, originally took a singular verb. In classical Latin the word has, in the singular, become an indeclinable adjective, so that Caesar writes *non longius mille et quingentis passibus* 'no farther than 1500 paces' (*Gallic War* I 22 1); the plural is still a noun, though it is used in apposition to another noun in such sentences as *extremum cornu claudebant quatuor milia mixti Cyrtii funditores et Elymaei sagittarii* 'the end of the flank was guarded by four thousand men, comprising Cyrtian slingers and Elymaean archers' (Livy XXXVII 40 9).

1.11   Latin expresses numbers from one thousand to nine hundred thousand by multiplying *milia*, as in *decem milia* 'ten thousand'. For one hundred thousand and above the multipliers were very often the adjectives *centena* and so on (see 2.5), so that, for example, Cicero expresses four hundred thousand as *quadragena milia* (*Pro Cluentio* 87).

1.12   To express one million and above Latin took the phrase *centena milia* 'one hundred thousand' and multiplied it using the numeral adverbs *decies* and so on (see 2.4); thus the Latin for one million is *decies centena milia* (literally 'ten times one hundred thousand'). But the *centena milia* was often dropped from these expressions, so that the numeral adverbs by themselves denote these high numbers, from *decies* 'one million' to *milies* 'one hundred million'. Cicero writes *uicies quinquies* 'two million five hundred thousand' (*In Verrem* I 92) and *ter et quadragies* 'four million three hundred thousand' (*Pro Flacco* 30).

1.13   For higher numbers than this Latin used the numeral adverbs as multipliers all over again; so Seneca has *quater milies* 'four hundred million' (*De Beneficiis* II 27 1), Cicero *septies milies* 'seven hundred million' (*Philippics* II 93) and Suetonius *quadringenties milies* 'forty thousand million' (*Vespasianus* 16 3).

CHAPTER TWO

OTHER NUMBER WORDS

2.1   The ordinal numbers are 'first, second, third' and so on. In Greek they form a series that begins πρῶτος δεύτερος τρίτος τέταρτος πέμπτος, and in Latin a series that begins *primus secundus tertius quartus quintus*. They are all adjectives. The Greek ones are declined like μόνος μόνη μόνον (but the feminine of δεύτερος is δευτέρα) and the Latin ones like *magnus magna magnum*.

2.2   Above 'nineteenth' the Greek ones end in –οστός and the Latin ones in –*esimus* (sometimes spelt –*ensimus*), as in τριακοστός 'thirtieth' and *octogesimus* 'eightieth'. Whereas English says 'twenty-sixth' both Greek and Latin tend to say 'twentieth-sixth' or 'sixth and twentieth'; so Thucydides writes δευτέρᾳ καὶ ἑξηκοστῇ ἡμέρᾳ 'on the sixty-second day' (I 108 2), Cicero has *annum iam tertium et uicesimum regnat* 'he is now in the twenty-third year of his reign' (*De Imperio* 7), and Livy refers to *censores uicesimi sexti* 'the twenty-sixth censors' (X 47 2). But sometimes in Greek and Latin, as in English, only the last term in the expression appears in the ordinal form; so Plato writes πεντεκαιεικοστός for 'twenty-fifth' (*Theaetetus* 175b 1), Herodotus εἷς καὶ τριηκοστός for 'thirty-first' (V 89 2) and Cicero *uno et uicesimo die* 'on the twenty-first day' (*Ad Familiares* XIV 5 1).

2.3   For the use of the ordinal numbers to express fractions see Chapter Five.

2.4   The numeral adverbs answer the question 'How many times?' (Greek ποσάκις, Latin *quot*). In Greek they comprise the series that begins ἅπαξ δίς τρίς τετράκις πεντάκις and in Latin the series that begins *semel bis ter quater quinquies*. Beyond this point the Greek ones all end in –άκις and the Latin ones in –*ies* (sometimes spelt –*iens*). Above 'twelve times' they are very rare in Greek but well attested in Latin. Their frequency in Latin is due to their important uses in expressing very large numbers (see 1.12) and in multiplication (see 6.3d).

2.5   Latin has a series of numeral adjectives that begins *singuli bini terni quaterni quini*; they are all declined like *magni magnae magna*. The adjectives in this series are in origin collectives and mean 'a group of one/two/three/four/five' and so on. In Latin literature they often have a distributive meaning 'one/two/three/four each'; but equally often they are merely synonyms for the cardinal numbers, as when Virgil describes Aegaeon as having *centum brachia centenasque manus* 'a hundred arms and a hundred hands' (not 'a hundred hands on each arm': *Aeneid* X 565-6). They are used especially as multipliers of *milia* (see 1.11). They frequently use *–um* instead of *–orum* as their genitive plural ending; so Cicero writes *pueri annorum senum septenumque denum* 'boys of sixteen and seventeen years' (*In Verrem* II 122).

2.6   By adding the suffix –αῖος to the ordinal numbers (see 2.1-2) Greek constructed a set of numeral adjectives of the form δευτεραῖος τριταῖος τεταρταῖος πεμπταῖος and so on; they are declined like λεῖος λεία λεῖον. As adjectives in agreement with the subject of a verb they provide grammatical alternatives to the dative of time when; so Thucydides writes διεφθείροντο οἱ πλεῖστοι ἐναταῖοι καὶ ἑβδομαῖοι 'the majority died on the ninth and the seventh days' (II 49 6), and elsewhere ὁδῷ δὲ τὰ ξυντομώτατα ἐξ Ἀβδήρων ἐς Ἴστρον ἀνὴρ εὔζωνος ἐνδεκαταῖος τελεῖ 'a fit man can complete the journey from Abdera to the Danube by the shortest route on the eleventh day' (II 97 1).

2.7   The Latin equivalent for πυρετὸς τριταῖος is *febris tertiana*. The Latin adjectives in *–anus* formed from the ordinal numbers (see 2.1-2) mean 'pertaining to the first/second/third' and so on. They appear infrequently in Latin literature, their commonest use being in the imperial period as designations for the men belonging to the various Roman legions: the *primani* were soldiers of the first (*pri-*

ma) legion and so on. The form *unetuicesimani* 'men of the twenty-first legion' appears in Tacitus (*Histories* II 43).

2.8   In Latin the adjectival ending *–arius* forms from the numeral adjectives (see 2.5) a set of words denoting size or content: *singularius* 'containing one', *binarius* 'containing two' and so on. They appear in such phrases as *fistula quinaria* 'a pipe with a diameter of five units' (see 13.1), *lex quinauicenaria* 'the 25 law' (which protected as minors those under the age of 25: see Plautus: *Pseudolus* 303) and *pater octogenarius* ' a father aged 80' (Pliny: *Letters* VI 33 2). To this set belongs *denarius*, originally *nummus denarius* 'a coin worth ten *asses*' (Varro: *On the Latin Language* V 173).

2.9   The termination *–arius* was added also to ordinal numbers (see 2.1-3). The resultant form meant 'pertaining to (the fraction referred to)'; so Petronius refers to the collectors of the 5% (*uicesima pars*) tax on the manumission of slaves as *uicesimarii* (*Satyricon* 65).

2.10   Added to the *uncia* fractions (see 5.9-10) the termination *–arius* formed another adjective, with similar meaning to the above, used especially in finance. A *heres semissarius* is one who inherits half (*semis*) of the estate, whereas a *heres unciarius* inherits only one twelfth (*uncia*). For the phrase *unciarium faenus*, quoted by Livy (VII 16 1 and 27 3) and Tacitus (*Annals* VI 16) from the Laws of the Twelve Tables, see 12.5.

2.11   There are instances in both Greek and Latin of numbers expressed by means of subtraction from a higher number rather than addition to a lower. Latin *duodeuiginti* 'eighteen' (literally 'two from twenty') and *undeuiginti* 'nineteen' (literally 'one from twenty') are of this type. There are forms like these (but not more than two is ever subtracted) in the cardinal number series in Latin up to *undecentum* 'ninety-nine', but they occur only rarely. Greek shows somewhat similar phrases, in which the number subtracted is not limited to two: Herodotus has ἐπ' ἔτεα δυῶν δέοντα εἴκοσι 'for eighteen years' (literally 'for twenty years lacking two': I 94 4), Demosthenes expresses 'twenty-nine' as τριάκονθ' ἑνὸς δέοντα (literally 'thirty lacking one': IX 23), and Thucydides expresses 'two hundred and ninety-two' as ὀκτὼ ἀποδέοντες τριακόσιοι (literally 'three hundred lacking eight': IV 38 5) and 'nine thousand seven hundred' as τριακοσίων ἀποδέοντα μύρια (literally 'ten thousand lacking three hundred': II 13 3). Ordinal numbers are sometimes similarly rendered: Thucydides, for example, writes ἑνὸς δέοντι τριακοστῷ ἔτει 'in the twenty-ninth year' (IV 102 3).

CHAPTER THREE

INCLUSIVE RECKONING; *ANTE* AND *POST*

3.1   The classical system now known as 'inclusive reckoning' is attested in Greek and Latin literature, particularly in the use of ordinal numbers for expressing dates and counting days. In such calculations the Greeks and Romans included the starting point in the reckoning.

3.2   The Greek use of inclusive reckoning is clearly shown by Demosthenes, who says that the twentieth of the month is the fifth day after the sixteenth and that the twenty-seventh is the fifth day after the twenty-third (XIX 59 and 60). This is why the standard gestation period of a human child was said to be ten months (τοὺς δέκα μῆνας: Herodotus VI 63 1). Herodotus uses a phrase διὰ τρίτου ἔτεος; the English translation of this is 'every second' (or 'every other') 'year', since τρίτος 'third' in the Greek phrase is by inclusive reckoning (II 4 1). By a similar process Xenophon's τῇ τρίτῃ means 'on the day after tomorrow' (*HG* III 1 17). One may also reckon back instead of forward: when Xenophon writes ἐχθὲς καὶ τρίτην ἡμέραν he means 'yesterday and the day before that' (*Cyropedia* VI 3 11).

3.3   The Roman use of inclusive reckoning is clear in dates (for which see further 14.13); to the Romans the twelfth of March was the fourth, not the third, day before the fifteenth (this explains the IV in the Roman abbreviation for 12th March: a. d. IV Id. Mar.). Celsus, like Demosthenes, draws attention to the use of inclusive reckoning when he writes *a septimo die undecimus non quartus sed quintus [est]* 'the eleventh day [is] not the fourth but the fifth after the seventh' (*On Medicine* III 4 15). The same author writes *tertio quoque die* for 'on every other day' or 'on alternate days' (*On Medicine* III 5 3).

11

3.4   To denote that the reckoning went forward or back the Romans used the adverbs *post* and *ante* respectively; so the phrase *tertio die post* means literally 'on the third day after' and, because it will normally involve inclusive reckoning, is to be translated 'on the second day after'. There was an alternative form of this phrase in which the *post* became a preposition and the phrase was written *post tertium diem*; this phrase too means 'on the second day after'. Livy uses both forms of the phrase (XLIII 11 7; XXXVIII 38 5). Similarly *ante quartum diem* means *quarto die ante* 'on the third day before' (literally 'on the fourth day before'). This type of expression is commonest in Latin dates, in which, for example, *ante diem quartum Idus Martias* means *quarto die ante Idus Martias* 'on the third (literally 'fourth') day before the Ides of March' (that is, March 12th: compare 3.3). Livy expresses the tenth, eleventh and twelfth of November as *ante diem quartum et tertium et pridie Idus Nouembres* (XLV 3 2).

3.5   The Latin phrase *tertio die post eum diem* means literally 'on the third day after that day' (and, by inclusive reckoning, will normally be translated 'on the second day after that day'). The same meaning might be expressed as *post tertium diem eius diei*, a curious use of the genitive which is found also in Caesar's phrase *postridie eius diei* 'on the day following that day' (*Gallic War* IV 13 4).

CHAPTER FOUR

SYMBOLS

4.1   Neither Greek nor Latin had a symbol for zero in its numeral system; this is because their systems did not use the concept of position, and so such a symbol was not needed. Ptolemy uses a symbol 0, perhaps a contraction of οὐδέν, in expressing the degrees of an angle using the sexagesimal system; the sign indicates a blank position, and in this respect its use is exactly parallel to the modern zero. But it occurs nowhere else in classical literature.

4.2   Of the two Greek systems of numerical symbols the acrophonic is the older. It is best attested in Attica, and the earliest extant example belongs to the middle of the 5th century BC. It was gradually superseded by the alphabetic system (see 4.3) and had become obsolete by about 100 BC. In the acrophonic system the symbol for 1 is I, and there are five other symbols consisting of the first letter of the Greek word for the number concerned (hence the name 'acrophonic'). So:

$$\begin{array}{ll}
\Pi \text{ (πέντε)} & = 5 \\
\Delta \text{ (δέκα)} & = 10 \\
H \text{ (ἑκατόν)} & = 100 \\
X \text{ (χιλιάς)} & = 1000 \\
M \text{ (μυριάς)} & = 10000
\end{array}$$

Symbols for 50, 500 and 5000 consist of Π with a small Δ, H or X tucked inside its top half. By this system 328 = ΗΗΗΔΔΠΙΙΙ.

4.3   Table 4.1 shows the Greek numerical symbols that make up the alphabetic system for writing numbers. In a system that does not use the concept of position 27 symbols are necessary in order to represent every number from 1 to 999; but the Ionic alphabet had only 24 letters. In order to make up the required number of symbols the arithmeticians added three archaic letters not used in ordinary writing, one in each group of nine. They were ς, a form of digamma, descended from Semitic *waw*, added in sixth position because it came sixth in the Phoenician alphabet; ϙ or koppa, added before r as its Phoenician equivalent *qoph* stood before *resh* in the Phoenician alphabet (but *qoph* was in nineteenth position therein); and ϡ, called *san* or *sampi*, descended from Phoenician *tsade*, added at the end to complete the tally (it came after *pe* = p in the Phoenician alphabet). In this system 365 = τξε and 8208 = ͵ησͺη. Alphabetic numerals often have a bar written above them, or may have an acute accent after them.

α	1	ι	10	ρ	100	͵α	1000
β	2	κ	20	σ	200	͵β	2000
γ	3	λ	30	τ	300	͵γ	3000
δ	4	μ	40	υ	400	͵δ	4000
ε	5	ν	50	φ	500	͵ε	5000
ς	6	ξ	60	χ	600	͵ς	6000
ζ	7	ο	70	ψ	700	͵ζ	7000
η	8	π	80	ω	800	͵η	8000
θ	9	ϙ	90	ϡ	900	͵θ	9000

Table 4.1: The Greek alphabetic system of numerals

4.4   M (μυριάδες) in front of a group of alphabetic numerals multiplies them by ten thousand, and MM (μύριαι μυριάδες) by one hundred million. So 1,324,580,000 = MMιγ.Μβυνη.

4.5   Table 4.2 sets out the Roman numeral system. This has some similarities to the Greek acrophonic system, in that C is the first letter of *centum* and M of *mille*; but these symbols may not have originated thus. By this system 78,692,484 is written D̄C̄C̄L̄X̄X̄X̄V̄Ī L̄X̄X̄X̄X̄Ī̄Ī CCCCLXXXIIII.

1	=	I
5	=	V
10	=	X
50	=	L
100	=	C
500	=	D
1000	=	M $\overline{\text{I}}$
5000	=	$\overline{\text{V}}$
10,000	=	$\overline{\text{X}}$
50,000	=	$\overline{\text{L}}$
100,000	=	$\overline{\text{C}}$ $\lceil\text{I}\rceil$
500,000	=	$\overline{\text{D}}$ $\lceil\text{V}\rceil$
1,000,000	=	$\overline{\text{M}}$ $\lceil\text{X}\rceil$
5,000,000	=	$\lceil\text{L}\rceil$
10,000,000	=	$\lceil\text{C}\rceil$
50,000,000	=	$\lceil\text{D}\rceil$
100,000,000	=	$\lceil\text{M}\rceil$

Table 4.2: Roman numerals

4.6   The use of subtractive forms such as IV = 4 and CM = 900 is later and less common than IIII and so on. The horizontal bar superscript (denoting *milia*) and the box circumscript (denoting *centena milia*) began to come into use towards the end of the republic. Earlier the symbol for 1000 was Φ (whence D for 500). Additional rings round this multiplied it by ten, so that 10,000 (as stylised for inscriptional purposes) was written ((I)) and 100,000 (((I))).

CHAPTER FIVE

FRACTIONS

A. GREEK

5.1 'Half'

(a) The prefix ἡμι- 'half', as in Homer's ἡμιτάλαντον χρυσοῦ 'a half-talent of gold' (*Iliad* XXIII 751), is of frequent occurrence.

(b) There is an adjective ἥμισυς 'half', as in Thucydides' τοῦ ἡμίσεος τείχους 'of half the wall' (II 78 2).

(c) The neuter singular of the adjective, τὸ ἥμισυ, is used from Homer on to mean 'half'. It governs a partitive genitive: so Thucydides writes τὸ ἥμισυ τοῦ στράτου 'half of the army' (IV 83 5). The word is occasionally assimilated to the number and gender of the dependent noun, as in Herodotus' τῶν Ἐχινάδων νησῶν τὰς ἡμισέας 'half of the Echinades islands' (II 10 3).

(d) The symbol C is often used (with numerals) for 'half'.

5.2 Other fractions with 1 as numerator.

(a) These are expressed as phrases with μέρος 'part' qualified by an ordinal adjective, as in Plato's τὸ πέμπτον μέρος τῶν ψήφων 'one fifth of the votes' (literally 'a fifth part of the votes': *Apology* 36b 1). Later writers may omit μέρος from such phrases, so that the ordinal numeral itself becomes a neuter noun denoting the fraction; so Diodorus Siculus expresses 'five and a quarter days' as πέντε ἡμέραι καὶ τέταρτον (I 50 2).

(b) There existed a series of compound neuter nouns of the type τριτημόριον 'third (part)', τεταρτημόριον 'fourth (part)' and so on;

17

Herodotus writes τεταρτημόριον τοῦ μισθώματος 'one fourth of the cost' (II 180 1). Aristotle has μυριοστημόριον 'one ten-thousandth' (*De Sensu* 445b 31).

(c) The symbol for such fractions is the alphabetic numeral followed by –ον (often reduced to "); so one quarter appears (with numerals) as δον or δ".

5.3   The neuter noun τὸ δίμοιρον 'two thirds' occurs as early as Aeschylus (*Supplices* 1070); it governs a partitive genitive. A symbol ω" is often used for 'two thirds'.

5.4   Fractions with numerator greater than 1.

(a) The literary treatment of such fractions is exemplified by Thucydides' very elegant phrase with μοῖρα, a synonym of μέρος: Πελοποννήσου τῶν πέντε τὰς δύο μοίρας 'two fifths of the Peloponnese' (I 10 2).

(b) Where possible, such fractions may be expressed as the sum of two or more fractions whose numerator is 1; thus Galen expresses 'one and two-thirds litres' as μίαν λίτραν καὶ ἡμίσειαν καὶ ἕκτον (literally 'one litre plus a half plus a sixth'), where ἡμίσειαν is an adjective and ἕκτον is a noun (*On the Synthesis of Drugs, by Types* I 13). Symbols may follow this system of addition, so that (for example) five sixths becomes Cγ" (a half plus a third).

(c) Mathematical writers in the later period use phrases similar to modern ones, expressing (for example) 'four fifths' as τέσσαρα πέμπτα (compare 5.2a). Symbols like modern fractions are sometimes used, except that the denominator is on top and the numerator underneath.

B. LATIN

5.5   'Half'

(a) The prefix *semi-*, equivalent to Greek ἡμι-, occurs frequently in words such as *semihora* 'half-hour'.

(b) There is a noun *semis semissis*, with masculine gender; it takes a partitive genitive, as in Pliny's *semissem Africae* 'half of Africa' (*Natural History* XVIII 35).

(c) Far more frequent in occurrence than *semis* is the noun *dimidium –i* (neuter) 'half'; it too takes a partitive genitive.

(d) The adjective *dimidius –a –um* was used before the Augustan age only in the phrase *dimidia pars*, synonymous with the noun *dimidium*; Cicero, for example, writes that the moon is *maior quam dimidia pars terrae* 'more than half (the size) of the earth' (*On the Nature of the Gods* II 103). Under the empire *dimidius* is used in agreement with any noun to mean 'half (of)', as in Juvenal's *dimidium crus* 'half a leg' (XIII 95).

5.6   Fractions with 1 as numerator were expressed by means of *pars partis* with an ordinal number in agreement, as in Caesar's *tertiam partem agri Sequani* 'a third part of the territory of the Sequani' (*Gallic War* I 31 10). For a quite different use of *pars* see 5.7 (c).

5.7   Fractions with numerator greater than 1.

(a) Such fractions are often expressed as the sum of two or more fractions whose numerator is 1; so Pliny writes *liquido patebit Europam totius terrae tertiam esse partem et octauam paulo amplius, Asiam uero quartam et quartam decimam, Africam autem quintam et insuper sexagesimam* 'it will be clearly apparent that

Europe comprises a third part of the whole land mass plus slightly more than one eighth, Asia a fourth plus a fourteenth, and Africa a fifth plus an additional sixtieth' (*Natural History* VI 210). One fifth plus one sixtieth is thirteen sixtieths; Pliny's mode of expression is arguably the simpler.

(b) In scientific and technical treatises phrases such as *quinta pars* are used in the plural with a cardinal number, as in Pliny's *deductis partibus nonis duobus* 'two ninths having been deducted' (*Natural History* XXXIV 159). *Partes* could be omitted from such phrases, as in the same author's *duae tertiae unius horae* 'two thirds of one hour', where *duae tertiae* has feminine gender because of the missing *partes* (*Natural History* VI 215).

(c) The Latin idiom for fractions whose numerator is one less than the denominator (such as four fifths, nine tenths) consisted of *partes* with a cardinal number equivalent to the numerator; thus four fifths is *quatuor partes* and nine tenths is *nouem partes*. Pliny uses the phrase *quinque partes lucri factae* for 'a saving of five sixths was made' (*Natural History* XXXIII 44). This type of expression is often accompanied by a phrase with ordinal number that adds in the final part and so accounts for the whole: a character in a play of Plautus says *decimam partem ei dedit, sibi nouem abstulit* 'he gave him one tenth and took nine tenths for himself' (*Bacchides* 667).

5.8   The Latin ordinal adjective *quotus* expects an ordinal number as its answer. Juvenal's question *quota portio faecis Achaei?* 'what fraction of our dregs is composed of Achaeans?' (*Satires* III 61) expects an answer of the form *tertia portio* (for *tertia pars*) 'a third part'. Questions such as Cicero's *quotus quisque est disertus?* (*Pro Plancio* 62) are an extension of this, being equivalent in meaning to *quota pars hominum est diserta?* 'what fraction of mankind is eloquent?'; the answer will be of the form *decimus quisque* 'each tenth man' (that is, one tenth, or one in ten). For the question *quota hora est?* 'what is the time?' see 14.17.

C. THE LATIN *UNCIA* FRACTIONS

5.9　　Latin possessed a set of duodecimal fractions with their own technical names as listed in Table 5.1. They were used especially in mensuration and finance. The fraction *sescuncia* 'one eighth' is also part of this system. It seems that Roman schoolboys learned to manipulate these fractions as if they were whole numbers. Horace comments that 'Roman boys learn to divide the *as* into lots of parts by lengthy calculations' and goes on to portray such an arithmetic class in session: "Tell us, Albinus' son: if you subtract *uncia* from *quincunx*, what is left? Come on, out with it!" "*Triens.*" "Good; you'll be able to look after your property. Add *uncia* [to *quincunx*] and what do you get?" "*Semis.*" (*Ars Poetica* 325-330).

Name	Symbol	Fraction of *as*	Unciae
as	I	(unity)	12
deunx	S==-	eleven twelfths	11
dextans	S==	five sixths	10
dodrans	S=-	three quarters	9
bes	S=	two thirds	8
septunx	S-	seven twelfths	7
semis	S	one half	6
quincunx	==-	five twelfths	5
triens	==	one third	4
quadrans	=-	one quarter	3
sextans	=	one sixth	2
uncia	-	one twelfth	1

Table 5.1: The Latin *uncia* fractions

5.10   *Uncia*, which functions as the unit in the above series, was itself subdivided as in Table 5.2. *Scrupulus* has an alternative form *scripulum* with neuter gender.

Name	Fraction of *uncia*	Fraction of *as*	Number of *scrupuli*
semuncia	one half	1/24	12
sicilicus	one quarter	1/48	6
sextula	one sixth	1/72	4
scrupulus	1/24	1/288	1

Table 5.2: Subdivisions of *uncia*.

5.11   Fractions not in the *uncia* series might be expressed by adding the appropriate terms in the series; thus one ninth might be expressed as *uncia binae sextulae* '1/12 + 2/72'. Compare 8.5.

D. SOME OTHER GREEK AND LATIN WORDS

5.12   Of very rare occurrence in Latin are the terms *libella* 'one tenth', *sembella* 'one twentieth' and *teruncius* 'one fortieth'. See further 11.8.

5.13   To express 'one and a half' Greek has the adjective ἡμι-όλιος, as in Herodotus' phrase τὰς περόνας ἡμιολίας ποιέεσθαι τοῦ κατεστεῶτος μέτρου 'to make the pins one and a half times as long as the standard measure' (V 88 2).

5.14   The Latin word *sesqui* 'one and a half' appears as an indeclinable noun once only, in the phrase *sesqui esse maiorem* 'to be one and a half times as big' (Cicero: *Orator* 188). Elsewhere it always appears as a prefix attached to a noun or adjective, as in Cato's *sesquipes* 'one and a half feet' (*On Agriculture* 46 1) and the corresponding adjective *sesquipedalis* 'measuring one and a half feet' (Caesar: *Gallic War* IV 17 3). Similarly, Plautus uses a noun *sesquiopus* 'one and a half days' work' (*Captiui* 725). The word *sescuncia* (see 5.9) is in origin **sesqui-uncia*.

5.15   The Herodotean phrase τρίτον ἡμιτάλαντον 'two and a half talents' means literally 'half a talent in third position' (that is, after the two whole talents: I 50 2). He also expresses 'four and a half cubits' as πέμπτη σπιθαμή (that is, 'one span in fifth place': see 7.3). Dio Cassius expresses 'seven and a half stades' as ὄγδοον ἡμιστάδιον 'a half stade in eighth position' (LIV 6). The Latin word *sestertius* is in origin an expression of the same type, being derived from **semistertius* 'a half in third position': it denoted the silver coin equivalent in value to two and a half *asses*.

5.16   Hippocrates expresses 'one and a half *kotylai*' as τρία ἡμι-κοτύλια, literally 'three half-kotylai' (*Muliebria* 34). Similarly, Plutarch expresses 'two and a half *mnai*' as πέντε ἡμιμναῖα (*Lycurgus* 12).

5.17   Both the Greek and the Roman whole number systems operate to the base ten, a fact recognised by Aristotle, who suggests that it may be because humans have ten fingers (see *Problems* XV 3: 910b 23 – 911a 4). The Roman *uncia* system of fractions is, however, duodecimal; and in astronomy sexagesimal fractions were used, as in the modern minutes and seconds (see 14.19-20). There is little trace of decimal fractions (but compare 5.12), and the type of fractional notation exemplified by modern 0.25 and 25% for one quarter was not known in the classical world.

CHAPTER SIX

SIMPLE ARITHMETIC

6.1   *Addition*. The usual Greek formulae are προστίθημι τρία τρίσιν 'I place three beside three' and συντίθημι τρία καὶ τρία 'I place together three and three'. Herodotus sets out an addition sum and introduces the total by means of the phrase τούτων πάντων συντιθεμένων 'all these being added up' (III 95 2). In Latin the usual formulae are *addo tria tribus* and *adicio tria tribus* 'I add three to three'; in either of these *ad tria* may be substituted for *tribus*.

6.2   *Subtraction*. The Greek verb is ἀφαιρῶ 'I remove', as in ἀφαιρῶ τρία ἀπὸ τῶν πέντε 'I subtract three from five'. The equivalent Latin verb is *demo*, as in *demo tria de quinque*; the bare dative or ablative case may appear instead of the phrase with *de*. For *addo* and *demo* as opposites compare Cicero's phrase *uerbo uno aut addito aut dempto* 'a single word being added or subtracted' (*De Oratore* II 109). In classical Latin *subduco* means, not 'I subtract', but 'I calculate'; but Cicero uses *deduco* for 'I subtract' (see 12.1).

6.3   *Multiplication*.

     (a) Greek adjectives of the form διπλάσιος τριπλάσιος τετραπλάσιος mean 'twofold threefold fourfold' or 'twice/three times/four times the size of'. They may be followed by a genitive of comparison. Plato has a series of expressions of this type at *Timaeus* 35b 4; it ends with a reference to a portion that is ἑπτακαιεικοσιπλασίαν τῆς πρώτης 'twenty-seven times the size of the first'. There is a generalised form πολλαπλάσιος, as in Herodotus' τὸ Ἑλλήνικον στράτευμα φαίνεται πολλαπλήσιον ἔσεσθαι τοῦ ἡμετέρου 'it seems that the Greek army will be many times the size of ours' (VII 48).

(b)   From πολλαπλάσιος comes a verb πολλαπλασιάζω 'I multiply', used by technical authors and late historians. The regular formula is τριάδα πεντάδι πολλαπλασιάζω 'I multiply three by five'.

(c)   In Latin the adjectives *duplex triplex* and so on mean 'twofold, threefold' or 'twice/three times the size of' and may be followed by a comparative construction: Pliny writes *duplex [ei] quam ceteris pretium* 'it is twice the price of the others' (*Natural History* XIX 9). The adjective *multiplex* is rare before the imperial period; Livy writes *multiplex quam pro numero damnum est* 'the loss is many times too large for the number involved' (literally 'many times larger than in proportion to the number': VII 8 1).

(d)   A verb *multiplico* is used, for instance, by Columella, who writes in his instructions for finding the area of a square *multiplicantur in se duo latera* 'two sides are multiplied together' (V 2 1). But the standard formula for multiplication tables in Latin used the numeral adverbs, as in *bis quina decem* 'twice five are ten'. The Latin for 'I multiply three by five' is *quinquies terna subduco* 'I reckon three five times' (the simple verb *duco* is also used in the sense 'I calculate'). Seneca has *nouem nouies multiplicata* 'nine multiplied by nine' (*Letters* LVIII 31).

(e)   The classical Latin equivalent of πολλαπλάσιος is *multis partibus maior*, as in Cicero's *sol multis partibus maior est quam terra* 'the sun is many times larger than the earth' (*On the Nature of the Gods* II 92), which would in Greek be ὁ ἥλιος πολλαπλάσιός ἐστι τῆς γῆς. Similarly τριπλάσιος = *tribus partibus maior*, and so on.

6.4   *Division.*

(a)   Whereas English divides 6 by 3 and gets the answer 2, both Greek and Latin divide 6 into 3 portions and get the same answer. Thus Plato defines an even number as ἀριθμὸς διαιρούμενος εἰς ἴσα δύο μέρη 'a number divisible into two equal parts' (*Laws* 895e

3), and Martianus Capella writes of the number 6: *hi et in singula diuidi possunt et in bina et in terna, cum et sexies singula et ter bina et bis terna fiant sex* 'these [that is, these six] can be divided into one, into two, and into three, since six ones, three twos, and two threes make six' (VII 753).

(b) Herodotus uses two accusatives (the 'whole-and-part' construction) with verbs of dividing, as in δυώδεκα μοίρας δασάμενοι Αἴγυπτον πᾶσαν 'dividing the whole of Egypt (into) twelve parts' (II 147 2). But Caesar writes, memorably, *Gallia est omnis diuisa in partes tres* 'all Gaul is divided into three parts' (*Gallic War* I 1 1).

6.5   *Square roots.* The following simple method of finding an approximation to the square root of a number that is not a perfect square uses a theorem of Euclid and may therefore have been in use in the classical world.

To find the square root of 200:
Take the square root of the next lower perfect square (196)   14
Subtract the perfect square from the number (200-196)   4
Take twice the square root of the perfect square (14x2)   28
Then the square root of 200 is approximately 14 4/28 (14 1/7)

6.6   *Counting on the fingers.* This was accepted practice. Herodotus writes: ἐπὶ δακτύλων συμβαλλόμενος τοὺς μῆνας, εἶπε ἀπομόσας Οὐκ ἂν ἐμὸς εἴη 'reckoning the months on his fingers he said with an oath: "The child cannot be mine"' (VI 63 2). Writing to Atticus on a matter of finance, Cicero remarks *hoc quid intersit, si tuos digitos noui, certe habes subductum* 'if I know your fingers you have already worked out the difference between these' (*Letters to Atticus* V 21 13). In the first of these passages a simple counting is all that is required; in the second the word is a metaphor for arithmetical skill.

6.7   *The abacus.* When faced with a calculation that could not be done on the fingers the ancient Greek or Roman drew a rough abacus and did the calculation on that. It consisted simply of a set of parallel vertical lines along which counters (such as pebbles or anything similar) could be moved. The lines could be drawn in a moment on a wax tablet or any suitable surface. The abacus uses the concept of position in exactly the same way as does the Arabic system of numerical notation: a counter placed in the right-hand column denotes 1, whereas the same counter placed three columns to the left denotes 1000. But the columns do not have to go up in the powers of ten: when Polybius writes 'these men are really like the pebbles on an abacus, which, according to the whim of the reckoner, have the value now of a *khalkous* and now of a *talanton*' (V 26 13) he envisages an abacus in which each column denotes one of the units of the Greek monetary system (in which one *talanton* = 288,000 *khalkoi*: see 11.3).

CHAPTER SEVEN

LINEAR MEASURES

7.1   Linear measures in the classical world were based on parts of the body, especially the arm, hand and foot. Greece and Rome both used the finger (δάκτυλος/*digitus*) and the foot (πούς/*pes*) as standard measures for short distances, reckoning the foot to be equivalent to sixteen fingers (that is, finger's breadths).

7.2   The earliest account of Greek linear measures appears in the text of Herodotus (II 149 3). From it Table 7.1 can be constructed. The ὄργυια 'fathom' is the distance between the finger-tips of a person with their arms stretched out sideways; the fingers (δάκτυλοι) appear in other Herodotean passages such as I 178 3, where he tells us that the royal cubit (used in ancient Persia and Egypt) was larger than the regular (μέτριος) one by three fingers (so it was 27 fingers long).

στάδιον 'stade'	=	6 plethra
πλέθρον 'plethron'	=	16 2/3 fathoms
ὄργυια 'fathom'	=	4 cubits
πῆχυς 'cubit'	=	1 1/2 feet
πούς 'foot'	=	4 palms
παλαστή 'palm'	=	4 fingers

Table 7.1: Greek linear measures from Herodotus

7.3   Other Greek measures occur less commonly. The πυγών of twenty fingers was the distance from the elbow to the bottom joint of the middle finger; it is found in Herodotus (II 175) and Xenophon (*Cynegeticus* 10 2). The σπιθαμή was half a cubit or twelve fingers; it was the distance between the tips of the thumb and little finger with the hand fully spread (sometimes called the greater span). It too appears in Herodotus (II 106). The λιχάς, equated to ten fingers, was the distance between the tips of the thumb and index finger with the hand fully spread (sometimes known as the lesser span); it is found in Pollux (*Onomastikon* II 158).

7.4   The Romans too had the cubit (*cubitus*), the foot (*pes*) and the finger (*digitus*), with the same numerical relationships (*rationes*) as in Table 7.1. But they also commonly divided their foot into twelve *unciae pedis* (see 5.9); and *uncia* in this context lived on to become the English word 'inch'.

7.5   The Romans measured longer distances by means of the *passus* 'pace'; this was what we would regard as a double pace (left foot to left foot). It was equivalent to five Roman feet. The word for a single pace was *gradus*. Long distances were measured in *milia passuum* 'thousands of paces', generally translated 'miles': thus *centena milia passuum* means 'one hundred miles' (the English word 'miles' comes from *milia* in these phrases). *Passuum* was sometimes omitted: Livy writes *Hannibal paucis post diebus sex milia a Placentia castra communiuit* 'a few days later Hannibal built a camp six miles from Placentia' (XXI 47 8).

7.6   Roman surveyors used a ten-foot rule known as *decempeda* and later as *pertica* (which has come into English as 'perch'). In late Greek the word ἄκαινα is used for this measure.

7.7   For longer distances Greek had the πλέθρον of one hundred feet and the στάδιον 'stade' of six hundred feet (compare Table 7.1). The παρασάγγης 'parasang' was a Persian, not a Greek, measure; the Greeks reckoned it as equivalent to thirty stades. Also Persian was the σταθμός or day's march, reckoned as approximately equivalent to five parasangs (Xenophon: *Anabasis* I 2 10); it represented the distance between the stations where the Persian king rested in his travels along the royal road.

7.8   In absolute terms the length of the foot might, and did, differ from place to place. Whereas the Attic foot was almost identical with the Roman foot at just under 300 mm, the Olympian foot measured approximately 320 mm. Tradition had it that the longer Olympian foot was the foot of Hercules, and Pythagoras was said to have deduced Hercules' stature using the ratio of the Olympian foot to that used elsewhere (Aulus Gellius: *Attic Nights* I 1). The other linear measures were in proportion to the foot, so that an Olympian stade of 600 Olympian feet was equivalent to 640 Attic feet, and so on.

7.9   Increasing contact between Greece and Rome made it necessary to find an equation linking the Roman mile and the Greek stade. If the stades are Attic stades of 600 Attic feet each, the mile is equivalent to 8 1/3 Attic stades, and we have the evidence of Strabo (VII 7 4) that Polybius used this equation. If, however, the stades are Olympian stades of 600 Olympian feet, then 7 13/16 are enough to make up the Roman mile; Plutarch writes that 'the mile is slightly less than eight stades' (*Gaius Gracchus* 7). This illustrates the sort of problems that arise when measures from one area are accommodated to those of another. These problems are considered at length by F. Hultsch in the Prolegomena to his Teubner text entitled *Scriptores Metrologici*.

7.10   Columella (V 1 6) and the elder Pliny (*Natural History* II 85) divide the mile into eight *Roman* stades of 625 *Roman* feet each. The Roman land surveyors had a measure called *actus*, equivalent to 120 Roman feet; see further 8.4.

7.11   When Egypt became a Roman province two systems of measurement were juxtaposed, the Roman and the Ptolemaic. The Ptolemaic foot was in proportion to the Roman as 6:5. This meant that the Ptolemaic cubit was equivalent to 1 1/2 Ptolemaic feet but 1 4/5 Roman feet. Table 7.2 presents some consequences of this.

	Ptolemaic cubit	Roman cubit
Roman feet	1 4/5	1 1/2
Ptolemaic feet	1 1/2	1 1/4
Roman palms	4 4/5	4
Ptolemaic palms	4	3 1/3
Roman fingers	19 1/5	16
Ptolemaic fingers	16	13 1/3

Table 7.2: Roman and Ptolemaic measures

Most of these figures are attested in late tables of measures (for details see Hultsch). At some stage, for ease of calculation, the Ptolemaic cubit was increased in size to make it equal to 2 Roman feet; this equation also is attested in late tables.

7.12   The modern equivalents in Table 7.3 are not intended to be perfectly accurate. The parasang measured approximately 5.33 km, and the *actus* 35.5 m.

	Olympian	Attic	Roman	Modern
1 mile			1.48 km	1.6 km
1 stade	192 m	177.4 m	185 m	
1 plethron	32 m	29.6 m		
1 fathom	1.92 m	1.8 m		
1 pace		74 cm	1.48 m	
1 cubit	480 mm	444 mm		
1 foot	320 mm	296 mm	296 mm	305 mm
1 inch			24.7 mm	25.4 mm
1 finger	20 mm	18.5 mm	18.5 mm	

Table 7.3: Modern metric equivalents of linear measures

CHAPTER EIGHT

MEASURING AREA AND VOLUME

8.1   It is not known who first multiplied units of length by units of breadth and obtained units of area; but Herodotus is probably right in thinking that the Greeks derived their knowledge of the techniques of measuring surface area from the Egyptians (Herodotus II 109 3).

8.2   Plato uses the adjectives τετράγωνος 'square' and ἐπίπεδος 'surface' and ἰσόπλευρος 'equal-sided' to describe numbers that are the product of two identical numbers (*Theaetetus* 147e-148a). But this terminology was not extended to the linear measures: the Greeks did not refer to 'square feet'. So when Socrates is questioning Meno's slave about the area of a square he asks: 'if this side measured two feet and this side two, how many feet (πόσων ποδῶν) would the whole be?' (*Meno* 82c 5): we would say 'how many square feet...'. Similarly, Hero of Alexandria finds the area of a square of side twelve feet to be 144 feet (*Geometrika* 5 2).

8.3   The Latin equivalent of ἐπίπεδος is *planus* (Aulus Gellius: *Attic Nights* I 20 1) and of τετράγωνος is *quadratus*. But when Columella multiplies feet by feet he may express the product either as *pedes quadrati* (V 1 6) or simply as *pedes* (V 2 5); Vitruvius (IX Praef. 4) uses simply *pedes*. At least one translator renders Pliny the Elder's *pedum quadratorum ternum* as 'three feet square', which is quite a different thing from 'three square feet' (*Natural History* XXXIII 75).

8.4   The Roman land surveyors used a standard measure called *actus*; it measured 120 feet. For area there was an *actus quadratus* of 120x120 feet, and the standard unit of land area had two of these side by side, making a rectangular figure with one side of 120 and one of 240 feet. This area was called *iugerum*; it was equivalent to

28,800 square feet. It divided up very conveniently using the *uncia* fractions (see 5.9-10) as in Table 8.1.

Name	Number of square feet	Approximate modern equivalent in hectares
iugerum	28,800	0.25
deunx iugeri	26,400	
dextans iugeri	24,000	
dodrans iugeri	21,600	
bes iugeri	19,200	
septunx iugeri	16,800	
semis iugeri	14,400	0.125
quincunx iugeri	12,000	
triens iugeri	9600	
quadrans iugeri	7200	
sextans iugeri	4800	
uncia iugeri	2400	
semuncia iugeri	1200	
sicilicus iugeri	600	
sextula iugeri	400	
scripulum iugeri	100	

Table 8.1: *Iugerum* and its subdivisions

8.5   Columella (V 2 2) expresses the area of a square field with sides of 100 feet as *triens iugeri* plus *sextula iugeri,* equivalent to 9600 + 400 square feet: compare 5.11.

8.6   Larger land measures were the *heredium* and the *centuria.* The *heredium* was an area of two *iugera*, equivalent to a square with side measuring two *actus.* The *centuria* was equivalent to one hundred *heredia* or two hundred *iugera*, and hence to a square with side measuring twenty *actus* or 2400 feet. According to Varro (*De*

Re Rustica I 10 2) there was also a measure known as a *saltus*, equivalent to 4 *centuriae*. The *centuria* was approximately equivalent to 50 hectares.

8.7 From Columella (V 2) we know how the area of other plane figures was calculated. The area of a square or a rectangle was found by multiplying base by height. The area of a right-angled triangle was half that of the rectangle formed by the two shorter sides. The area of an equilateral triangle is found by adding together one third and one tenth of the square on one side. The area of a circle is found by squaring the diameter, multiplying by eleven, and dividing by fourteen. The fraction 11/14 is the approximate relation established by Archimedes between the area of a circle and the square on its diameter (*Measurement of the Circle* 2); it is equivalent to a value of 22/7 for π.

8.8 The Greek word for 'solid' or 'three-dimensional' was στερεός; at *Theaetetus* 148b 2 Plato uses τὰ στερεά to refer to what would now be called cubes and cube roots. The Latin equivalent of στερεός is *solidus*. Vitruvius (IX Pref. 13) refers to the problem of constructing an altar with twice the volume of a given altar ("duplication of the cube"). The volume of the right-angled cube, prism or cylinder could be found by calculating the area of the end and then multiplying by the height. It took a thinker of the calibre of Archimedes to realise that the volume of any solid object, however irregular, could be found by measuring the amount of water displaced by it; Vitruvius tells the story, with much admiration, at IX Praef. 9-12).

CHAPTER NINE

MEASURING WEIGHT

9.1   The principal weight standards of ancient Greece were the Aeginetic and the Euboic; the latter became the Attic standard when Solon (according to tradition) introduced it to Attica. The Attic-Euboic standard was approximately 70% of the Aeginetic, which it replaced in most areas after the age of Alexander.

9.2   The principal Greek units of weight were τάλαντον 'talent', μνᾶ 'mina', δραχμή 'drachma' and ὀβολός 'obol'. Table 9.1 shows their relationships.

	minas	drachmas	obols
talent	60	6000	
mina		100	600
drachma			6

Table 9.1: Greek weights

On the Attic-Euboic standard the drachma weighed approximately 4.33 grams.

9.3   The standard weight unit of Rome was *libra* 'pound'. The word was sometimes accompanied by *pondo*, the phrase *libra pondo* meaning literally 'a pound in weight'. But *libra* was often conventionally dropped from this phrase, so that *pondo* by itself came to be regarded and used as an indeclinable noun meaning 'pound(s) weight', as in Cicero's *auri quinque pondo abstulit* 'he stole five pounds of gold' (*Pro Cluentio* 179). The English word 'pound' comes from this indeclinable *pondo*, but the abbreviation 'lb' is from *libra*.

37

9.4   The pound was divided by means of the *uncia* fractions (see 5.9-10), so that *uncia librae* is one twelfth of a pound; the English word 'ounce' comes from *uncia* in this phrase. English 'scruple' comes from *scrupulus librae* or *scripulum librae* '1/288 of a pound' (equivalent to 1/24 of an ounce). So Martial says that, when Marulla measures the weight of something, *libras scripula sextulasque dicit* 'she gives it in pounds, sixths of an ounce and 24ths of an ounce' (X 55 3). In modern terms, one *uncia librae* is equivalent to approximately 27.3 grams and one *libra* to 327.45 grams.

9.5   Terms imposed by the Romans on Antiochus III in 188 BC include the statement "the talent is to weigh not less than eighty Roman pounds" (Polybius XXI 43 19). This makes the mina equivalent to one and one third pounds and the pound to seventy-five drachmas.

9.6   Greek authors from Polybius on use λίτρα to translate Latin *libra*. Later Greek authors use γράμμα to translate *scrupulus (librae)*. These Greek words are the source of English (French) 'litre' and 'gramme'.

9.7   The Romans sometimes used their denarius coin (see 11.7ff) as a measure of weight. As a unit of weight the denarius decreased in size due to inflation.

9.8   From c. 211 BC to c. 187 BC it weighed *sextula unciae* ('one sixth of an ounce') or four *scrupuli* (about 4.5 grams); in other words, 72 were made from one *libra* of metal. The denarius was therefore very close to the weight of the Attic drachma (about 4.33 grams: see 9.2).

9.9   From c. 187 BC to AD 64 the denarius weighed one seventh of an ounce (that is, 84 were minted from a *libra* of metal); so Celsus writes *in uncia pondus denarium septem* 'in the *uncia* there is a weight of seven *denarii*' (*On Medicine* V 17 1C).

9.10   In AD 64 Nero reduced the weight of the denarius to one eighth of an ounce or three *scripula*. It now weighed approximately 3.375 g, and 96 were minted from one *libra* of metal. After Nero's time the denarius weight began to be referred to as an Attic drachma; but this is not the same weight as the drachma of ancient Greece (compare 9.8). At this stage the following table of Roman weights applies.

	unciae	denarii	scripula
libra	12	96	288
uncia		8	24
denarius			3

Table 9.2: Roman weights after Nero.

9.11   When the denarius weighed 3 *scrupuli* the weight known to the Romans as *unciae dimidius scrupulus* was one sixth of a denarius, the same relation as the obol to the drachma in Greek; but at approximately 562.5 mg it was smaller than the obol.

9.12   In c. AD 312 the emperor Constantine introduced a new gold coin, the *solidus*, weighing one sixth of an ounce (the same weight as the original *denarius*). The weight equivalent to 1/24 of this was named *siliqua*. The Greek word used to translate *siliqua* was κεράτιον, from which modern 'carat' is derived.

CHAPTER TEN

MEASURING CAPACITY

10.1   The terms most commonly used in classical Greek to measure liquid capacity are derived from the names of containers and receptacles. The basic unit is the κοτύλη 'cup', equivalent to 6 κύαθοι 'ladles'. There are multiples called χοῦς 'pitcher' and μετρητής 'measurer', as in Table 10.1.

	χόες	κοτύλαι	κύαθοι
μετρητής	12	144	864
χοῦς		12	72
κοτύλη			6

Table 10.1: Greek measures of liquid capacity

10.2   The absolute size of the κοτύλη varied somewhat from place to place; the commonest sizes are (in modern terms) 240 ml and 270 ml.

10.3   Herodotus refers to a golden bowl χωρέων ἀμφορέας ἑξακοσίους 'with a capacity of six hundred amphorae' (I 51 2); the ἀμφορεύς as a measure of capacity was perhaps the same as the μετρητής. The large earthenware jar called a κεράμιον may also have been used as a measure denoting this capacity.

10.4   In later Greek the ὀξύβαφον 'saucer' is equated to one quarter of a κοτύλη. This and the χήμη 'clam shell' appear in the Hippocratic corpus but may not have had their size fixed at that stage.

10.5   Roman measures of liquid capacity under the republic were based on the *quadrantal*, equivalent to one cubic foot or approximately 26 litres. Its subdivisions appear in Table 10.2. The

41

relationships between *congius, sextarius* and *hemina* can be deduced from Cato's recommendations concerning the wine ration for farm hands in *Agriculture* 57; reckoning a month at 30 days he says *(bibant) heminas in dies, id est in mense congios IIS* '(let them drink) a *hemina* per day, that is, 2 1/2 congii per month', and, later in the year, *(bibant) in dies sextarios, id est in mense congios quinque* '(let them drink) a *sextarius* per day, that is five *congii* per month'. But his *amphora* is slightly smaller than the later one: at another time of the year *(bibant) in dies heminas ternas, id est in mense amphoram* '(let them drink) three *heminae*, that is, an *amphora* per month' (ninety *heminae*, as contrasted with the later 96).

	congii	sextarii	heminae
quadrantal	8	48	96
congius		6	12
sextarius			2

Table 10.2: Roman measures of liquid capacity

10.6   More measures were added later, and Greek equivalents were established for the various Latin terms. The system under the empire was as follows; the modern equivalents are approximate.

quadrantal	2 urnae	26 litres
urna	4 congii	13 litres
congius (χοῦς)	6 sextarii	3.25 litres
sextarius (ξεστής)	2 heminae	540 ml
hemina (κοτύλη)	4 acetabula	270 ml
acetabulum (ὀξύβαφον)	1 1/2 cyathi	67.5 ml
cyathus (κύαθος)	4 cochlearia	45 ml
cochlear (χήμη)		11 ml

Table 10.3 Roman measures of liquid capacity

The Roman *amphora* was equivalent to the *quadrantal* at 8 congii. The Greek amphora was equivalent to the μετρηρτής at 12 χόες. Hence the Greek amphora was one and a half times the size of the Roman one (approximately 39 litres).

10.7   The elder Pliny writes: *cum acetabuli mensura dicitur, significat heminae quartam* 'when the *acetabulum* is used as a measure it signifies a quarter of a *hemina*' (*Natural History* XXI 185). The other relationships are from late legal and technical authors.

10.8   We have the evidence of Galen that the Romans divided the *hemina* into twelve parts called, as always, *unciae* (see 5.9ff), and that the measure consisting of these twelve *unciae*, the *hemina*, was sometimes known by analogy as a *libra*. This measure was used especially for measuring oil. Because Galen writes in Greek he calls it λίτρα (e.g. *On Maintaining Health* IV 8: Hultsch 30); there was an alternative form λίτρον from which modern 'litre' is derived. It must be stressed that neither the measure called *libra* nor its contents weighed a pound (*libra*); it was called that purely because it contained twelve *unciae* (which are the forerunners of modern fluid ounces). Galen is careful to distinguish between 'metric' *unciae* and *librae* (for measuring capacity) and 'stathmic' *unciae* and *librae* (for measuring weight). Speaking of Rome, he writes: ἐπιχώρια γὰρ ταῦτα ὀνόματα τό τῆς λίτρας καὶ τὸ τοῦ ξέστου καὶ τὸ τῆς οὐγγίας 'the local names are *libra*, *sextarius* and *uncia*' (*On the Synthesis of Drugs, by Types* I: Hultsch 34).

10.9   In both Greece and Rome most measures of liquid capacity were used also for dry capacity. Thucydides, for example, has the κοτύλη as both a liquid and a dry measure: κοτύλην ὕδατος καὶ δύο κοτύλας σίτου 'a κοτύλη of wine and two κοτύλαι of food' (VII 87 2). However, there were a few terms used solely for dry measure.

10.10   For dry measures Greek had the χοῖνιξ, equivalent to four κοτύλαι (just over 1 litre), and the μέδιμνος of 48 χοίνικες (approximately 52 litres). The Romans borrowed the μέδιμνος as *medimnus* and subdivided it into 6 *modii*, and this meant that the *modius* was equivalent to sixteen *sextarii* (just over 8.5 litres).

10.11   The capacity of liquid and solid measures was frequently defined in terms of the weight of some substance which they would hold; this is similar to the modern concept of the density of liquids. The standard figure in Rome was that one *hemina* would hold ten (stathmic) *unciae* of water or wine but only nine (stathmic) *unciae* of oil; which meant that wine or water was heavier than oil by one ninth. The *Leges publicae* have the equation *congius uini decem pondo siet* 'the congius of wine is to weigh ten pounds'. It is to be observed that a metric *libra* of water (see 10.8), equivalent to twelve metric *unciae*, weighed, not a stathmic *libra*, but ten stathmic *unciae*.

10.12   The *hemina* (and κοτύλη), weighing 10 *unciae*, therefore weighed 80 Attic drachmas (see 9.10). Galen found references in the literature to an Attic κοτύλη weighing 60 instead of 80 Attic drachmas; being therefore 3/4 the size of the regular *hemina*/κοτύλη it was equivalent to nine metric *unciae*. Galen refers at one point to 'the usual problem' (τὴν συνήθη ἀμφιβολίαν: *On the Synthesis of Drugs, by Types* VI 13, see Hultsch 50) whether the κοτύλη = 9 or 12 metric *unciae*. This smaller κοτύλη appears to have arisen from a misapprehension by someone who confused the two Attic drachmas (see 9.8 and 9.10).

CHAPTER ELEVEN

MEASURING VALUE

11.1   Value is measured by means of coins, which originated in standard weights of precious metals. An earlier system of measurement, in which the ox was the unit of value, is attested in Homer (where the adjective ἐννεάβοιος means 'worth nine oxen': *Iliad* VI 236) and in the Latin word *pecunia* 'money' (from *pecus* 'cattle': Pliny: *Natural History* XVIII 11).

11.2   The Greek units were the talent, mina, drachma and obol (see 9.2). Each referred to that weight of silver, and only the last two were small enough to function as coins.

11.3   Coins representing multiples and fractions of the drachma and obol were minted at various times and places, including especially the tetradrachm, didrachm, triobol (= 1/2 drachma) and hemiobol. Coins down to 1/8 obol were issued during the fourth century BC; but a silver coin worth only 1/8 obol was inconveniently small and 'was superseded by its equivalent in bronze, the *khalkous*, when that metal came into general use, probably after the middle of the fourth century' (Head).

11.4   Each state that issued coinage chose one denomination as the coin of standard weight or στατήρ 'stater'; its weight was fixed absolutely, and the weights of the other coins in the system were fractions or multiples of it. Sums of money were, however, usually reckoned in drachmas.

11.5   The two oldest weight standards used in the Greek world for issuing coinage are known as the Aeginetic standard and the Euboic standard. The Aeginetic standard was used for the first silver coinage of ancient Greece; the stater was a didrachm weighing between 12.25 and 12.35 g. When (according to tradition) Solon introduced coinage to Attica he used the lighter Euboic standard, and Athenian coinage is said to follow the Euboic-Attic weight standard. The first Athenian coins struck were didrachms weighing approximately 8.5 g, and there was a drachma weighing half this. The Athenian coinage using the Euboic-Attic standard was eventually adopted by Alexander the Great and so became the most important in the ancient world. Before his time other standards were in use in other areas.

11.6   The Romans began to issue coins early in the third century BC. This bronze coinage was based on the *as*, which originally weighed one *libra* and was subdivided by means of the *uncia* fractions (see 5.9) into *semis, triens, quadrans, sextans, uncia* and *semuncia*. After the introduction of the denarius coinage in c. 211 BC the bronze coinage was issued on a sextantal standard: that is, the weight of the *as* was reduced from one pound (*libra*) to 1/6 pound (*sextans librae*; but the *as* and its subdivisions retained their old names). Later the *as* was further reduced in weight to *uncia* (c. 155 BC) and to *semuncia* (91 BC). From this last date the *quadrans* was the smallest coin minted.

11.7   According to Pliny the Elder (*Natural History* XXXIII 44) the first Roman silver coins were issued in 269 BC; they were didrachms. In c. 211 BC they were replaced by the famous *denarius* coinage, consisting of a *denarius* equivalent in value to 10 *asses* (hence its name and its symbol X), a *quinarius* equivalent to 5 *asses* (symbol V), and a *sestertius* equivalent to 2 1/2 *asses (symbol IIS)*. The *denarius* weighed 4 scruples or 1/6 *uncia* (compare 5.10).

11.8   There was also a silver coin worth one *as* and called *libella*: Varro writes *nummi denarii decuma libella* 'the *libella* is one tenth of a denarius' (*On the Latin Language* V 174). The coin is perhaps identical with the otherwise unknown *nummus assarius* to which he refers elsewhere. It was subdivided into halves called *sembellae* and quarters called *teruncii*. The *libella* coin and its subdivisions were not minted after the Second Punic War; but the words retained their meanings as fraction-words, *libella* denoting one tenth, *sembella* one twentieth and *teruncius* one fortieth (see 5.12). It is in this sense that Pliny the Elder writes *etiam nunc libella dicitur* 'the word *libella* is still in use' (*Natural History* XXXIII 42). This was so particularly in statements of inheritances (see 12.10).

11.9   The denarius was minted for a period of several centuries. The quinarius went out of issue in the 170s BC and was revived in 101 BC and later; the revived *quinarius* is known as a *uictoriatus* from the design on the reverse. The *sestertius* went out of issue after only a few years, but was briefly revived c. 91 BC (and see also 11.11).

11.10   In c. 155 BC the weight of the denarius was decreased to 1/7 *uncia* (see 9.9), and in c. 123 BC (or perhaps as early as c. 141 BC) the denarius was reassessed at 16 *asses* instead of ten. Its symbol was changed, briefly to XVI, then to X̶. The quinarius and sestertius, when minted, continued to be one half and one quarter of a denarius (therefore now 8 and 4 *asses* respectively), symbolised V̶ and H̶S̶. The latter is nowadays usually printed for convenience HS. Before the reassessment of the denarius all sums of money were officially reckoned in *asses*; after it they were reckoned in *sestertii*, even though that coin was rarely minted.

11.11   The Roman coinage system inaugurated by Augustus in c. 23 BC was a development from the republican system outlined above. It had as its basis four different coins made of four different metals: an *as* made of copper (not bronze), a *sestertius* worth 4 *asses* and made of orichalcum (an alloy of copper and zinc), a silver *denarius* worth four *sestertii*, and a gold *aureus* worth 25 *denarii*. The *denarius* weighed 1/7 *uncia*, so that there were 84 to the *libra* (compare 9.9); the *aureus* was twice the weight of this (42 to the libra) but 25 times the value.

11.12   Other Augustan coins were the *dupondius* (made of orichalcum and worth 2 *asses*) and the *quadrans* (made of copper and
worth 1/4 *as*). The coins most likely to be handled by ordinary
people were the *dupondius* and the *as*.

11.13   In Petronius' *Satyricon* the adjectives *sestertiarius* and *du-
pondiarius* (compare 2.8) are used to mean 'almost worthless' (45 8
and11; 58 5; 74 15).

CHAPTER TWELVE

SOME FINANCIAL MATTERS

12.1 *Bookkeeping.* There is some useful technical vocabulary in a passage of Cicero's *On Duties* (I 18 59). We must take account, he says, of our moral behaviour *ut boni ratiocinatores officiorum esse possimus et addendo deducendoque uidere quae reliqui summa fiat, ex quo quantum cuique debeatur intellegas* 'so that we may become good book-keepers of duties and, by adding and subtracting, may see what the total remaining is, for it is from this that one may discover how much one owes to each person'. Catullus, however, has quite a different aim in his kissings of Lesbia (V 10-11): *dein, cum milia multa fecerimus,/ conturbabimus illa ne sciamus* 'then, when we have kissed many thousands of times, we shall confuse the reckoning so that we shall not know (how many)'.

12.2 *Moneylending* is a topic of major interest in both Greek and Latin literature. In Greek the phrase δανείζειν ἐπὶ τόκῳ (Plato: *Laws* 742c) means 'to lend out money at interest'; the principal is τὸ ἀρχαῖον or τὸ κεφαλαῖον, interest is ὁ τόκος, and compound interest is τόκοι τόκων ('interest on interest': Aristophanes: *Clouds* 1156). In Latin the verb *faenerari* means 'to lend at interest'; the principal is *caput* or *sors* and interest is *faenus* or *usura* (often plural: *usurae*). So Marcus Manlius exhorts the senate: *sortem reliquam ferte; de capite deducite quod usuris pernumeratum est* 'take back the rest of the debt; deduct from the principal what has been paid in interest' (Livy VI 15 10).

12.3 In both Greek and Latin the amount of interest is expressed as a monthly rather than an annual rate: the borrower paid the lender so much a month for the use of the money. Interest was payable (or the loan repayable plus interest) in Greece on the last day of the month and in Rome on the first day of the next month.

12.4   In Greek the rate of interest was expressed by means of a prepositional phrase with ἐπί. A basic rate was ἐπὶ δραχμῇ, which meant that the borrower paid the lender one drachma per month for each mina borrowed. As 1 mina = 100 drachmas (see 9.2) this is equivalent to 1% per month or 12% per annum. Thus a person who lends money ἐπ' ὀκτὼ ὀβολοῖς is charging one and one third per cent per month or 16% per annum.

12.5   In Latin a rate of 1% per month is expressed as *centesima* (sc. *parte*). Higher rates were denoted by multiplying the *centesimae*: when Cicero writes *binis centesimis faeneratus est* he refers to a rate of 2% per month or 24% per annum (*In Verrem* III 165). There is an extortionate rate of *quinae* (60% per annum) at Horace: *Satires* I 2 14. Rates less than 1% per month might be expressed in *unciae* (that is, twelfths of the basic rate of 1% per month), so that a rate of *uncia* would be equivalent to 1% per annum. But the phrase *unciarium faenus* used by Livy (VII 19 5) and Tacitus (*Annals* VI 16) probably means that the annual interest payment amounts to one twelfth of the principal, an annual rate of 8 1/3 per cent; and *semunciarium faenus* is half that or 4 1/6 per cent per annum.

12.6   If interest was not paid when it fell due it might be added to the principal and attract interest at the same rate. For this system, known to us as compound interest, Cicero (*Letters to Atticus* V 21 11) uses a Greek word ἀνατοκισμός which is not attested in classical Greek literature (and compare 12.2 above).

12.7   For terms other than one month the interest rate might be expressed in Greek as that fraction of the principal which the borrower must pay in addition to the principal at the termination of the loan. Thus ὁ ἐπιδέκατος τόκος 'interest at an additional tenth' meant that the borrower paid back when the time came the amount of the principal plus an additional tenth of that amount as interest. Aristotle refers to τόκοι ἐπίτριτοι 'interest at an additional third' (*Rhetorica* 1411a 17). For a term of one year, ὁ ἐπιδέκατος τόκος is equivalent to a monthly rate ἐπι πέντε ὀβόλοις (10% per annum: compare 12.4).

12.8   Among ways of losing your money Martial mentions *debitor usuram pariter sortemque negabit* 'your debtor will repudiate interest and principal alike' (V 42 3). Another item in his depressing list runs *mercibus extructas obruet unda rates* 'a wave will overwhelm your ships loaded with merchandise' (V 42 6). For the latter contingency, however, the ancients had a sort of insurance policy now known as bottomry.

12.9   *Bottomry.* This arrangement was regarded by the ship-owner as insurance and by the insurer as a short-term, high-risk investment attracting high rates of interest. It was common in both Greece and Rome. In essence it was a loan to the shipowner on the security of the ship and/or its cargo. The shipowner would approach a moneylender, who would advance a sum of money at an agreed rate of interest. The ship then sailed with its cargo. If it reached port in safety the shipowner would pay back to the moneylender the amount of the loan plus interest; but if the ship and cargo were lost on the voyage the shipowner kept the money and the lender got nothing. The term of the loan was the end of the voyage; payment was required within twenty days of the ship's safe arrival. In one of Demosthenes' speeches (50.17) the interest in such a transaction is ἐπόγδοος 'an additional eighth' on a loan of 800 drachmas: the moneylender will get back 900 drachmas if the ship arrives safely.

12.10   *Inheritances.* In Latin the *uncia* fractions (see 5.9) are used to denote the portion of a deceased estate which someone inherits; the grammatical form is a prepositional phrase consisting of *ex* + ablative. The phrase *heres ex asse* denotes the sole heir to an estate, *as* being the word for the whole which is split up into the *uncia* fractions. Cicero writes concerning a deceased estate *Caesar ex uncia...sed Lepta ex triente*, which means that Caesar gets one twelfth of the estate and Lepta one third (*Letters to Atticus* XIII 48

1). A different type of expression appears in a clause of a will quoted by Horace: *quartae sit partis Ulixes...heres* 'let Ulixes be heir of the fourth part' (*Satires* II 5 100-1): here *quartae partis* is equivalent to *ex quadrante* and refers to a quarter of the estate.

12.11   *Libella* and *teruncius* (see 11.8) are used in this context: Cicero writes *fecit palam te ex libella me ex teruncio* 'he openly left you one tenth of his estate and me one fortieth' (*Letters to Atticus* VII 2 3).

12.12   *Wagers.* The phrase *pignus dare cum* (+ ablative) means 'to make a bet with (someone)'. If one was certain that something was the case one challenged one's opponent with the words 'bet me that this is not true' *pignus da mihi ni hoc uerum sit*. The stake was expressed by means of a prepositional phrase consisting of *in* + accusative; it might be a sum of money, a jar of wine, or even a kiss. In one Plautine passage two stakes are mentioned: *ni ergo matris filia est,/ in meum nummum in tuom talentum pignus da* 'bet me your talent to my coin that she is not a mother's daughter' (*Epidicus* 700-1). Here something like the modern concept of odds is involved; if the coin is, as often, a didrachm, the odds are 3000 to 1.

CHAPTER THIRTEEN

SIZES OF PIPES AND NOZZLES

13.1   It is mainly because of Frontinus' treatise *On Aqueducts* that we know how the Romans calculated and expressed the sizes of pipes used in their water reticulation system. The Latin word for 'pipe' is *fistula*, and the size was expressed by numerical adjectives with the termination *–aria* (compare 2.8). Thus *fistula quinaria* means 'size five pipe'. There was a special word for the bronze nozzle that regulated the amount of supply to the individual householder: this word is *calix*. The nozzle is sometimes referred to as *modulus* 'meter'.

13.2   In his treatise *On Architecture* Vitruvius gives instructions for the manufacture of lead pipes (*fistulae*) and says that 'when a pipe is made from a sheet of lead 50 fingers wide it is called a size 50 pipe (*fistula quinquagenaria*)' (VIII 6 4). These fingers are finger's breadths (see 7.1). Frontinus criticises this system as being not sufficiently accurate, because when the sheet of lead is bent round to make the pipe 'it is contracted on the inside and stretched on the outside' (*Aqueducts* I 25). There were ten sizes of lead pipe, running from size 5 to size 100.

13.3   Frontinus has a great deal to say about the sizes of the bronze nozzles, because householders who received the supply were taxed on the size of these. The sizes run from 5 to 120. The system was in two parts, for smaller and larger sizes respectively.

13.4   For nozzles up to size 20 the internal diameter is measured in units equivalent to 1/4 of a finger (that is, of a finger's breadth), and the number of units is the size of the nozzle. The *quinaria* is the smallest nozzle; it has an internal diameter of 5 units = 1 1/4 fingers or approximately 2.3 cm.

13.5   The larger sizes start with size 20; their size number is the area of the orifice in square fingers, and the diameter was calculated by working back from the area. The size 20 nozzle (*uicenaria*) has an orifice whose area is 20 square fingers; its internal diameter is 5 *digiti* plus 13 *scripula digiti* (see 5.10). The sizes go up to 120 (*calix centenum uicenum*: for the genitive plural *centenum uicenum* see 2.5), whose internal diameter was 12 *digiti* plus 102 *scripula digiti*. But Frontinus points out (I 63) that for this size a larger diameter of 16 *digiti* was used in practice; and similarly (I 62) that the diameter of the size 100 nozzle, which was correctly 11 *digiti* plus 81 *scripula digiti*, was in practice rounded up to 12 *digiti*. Table 13.1 gives the theoretical sizes of the nozzles.

Size	Latin name	Diameter in digiti	Area in sq. dig.	Capacity in quinariae
5	quinaria	1 + quadrans		1
8	octonaria	2		2 + 161 scr.
12	duodenaria	3		5 + 219 scr.
20	uicenaria	5 + 13 scr.	20	16 + 84 scr.
40	quadragenaria	7 + 39 scr.	40	32 + 168 scr.
60	sexagenaria	8 + 212 scr.	60	48 + 251 scr.
80	octogenaria	10 + 20 scr.	80	65 + sextans
100	centenaria	11+ 81 scr.	100	81 + 130 scr.
120	centenum uicenum	12 + 102 scr.	120	97 + dodrans

Table 13.1: Water supply nozzles

13.6   The capacity of a nozzle was reckoned in units also called *quinariae*. One *quinaria* was defined as the capacity of the size 5 nozzle; and the figures for the capacities of the other nozzles represent the number of times by which the area of their orifice is greater than that of the size 5 nozzle. This means that the capacities in *quinariae* are proportional to the squares on the diameters of the various pipes: Frontinus gives the capacity of the size 80 nozzle as 65 *quinariae* plus *sextans quinariae* (see 5.9 and Table 13.1).

13.7   Frontinus' figures are given in full in *Aqueducts* I 39-63; they provide a virtuoso display of calculation using the *uncia* fractions.

CHAPTER FOURTEEN

MEASURING TIME

14.1   The natural sources from which the Greeks and Romans derived their measures of time are three: the cycle of the seasons ('year'), the phases of the moon ('month'), and the alternation of light and darkness ('day'). Establishing the arithmetical relations between these was a long-lasting problem.

A. YEARS

14.2   It was in the middle of the fifth century BC that the Greek astronomers established that the length of the year is 365 days plus a fraction of a day. During the second century BC Hipparchus refined this figure to 365 1/4 days less 1/300 of a day. The calendar worked out by Sosigenes for Julius Caesar ignored the 1/300 of a day and therefore needed correction in the sixteenth century AD.

14.3   Both the Greeks and the Romans designated years with reference to the magistrates holding office. In Rome the year was named after one or both consuls: Horace's line (*Odes* III 21 1) *O nata mecum consule Manlio* 'O [jar of wine], born with me when Manlius was consul' designates the year which we would call 65 BC. In Athens the ἄρχων ἐπώνυμος gave his name to the year. In Rome (from 153 BC on) the magistrates took office on 1 January; in Athens they did so on 1 Hecatombaion, which was in midsummer. Hence each Greek year covered half each of two consecutive Roman years and vice versa.

14.4   A system of designating years by consecutive counting did not appear until Eratosthenes of Cyrene (c. 275-194 BC) worked one out using the four-year cycle of the Olympic Games. After his time all Greek chronology was based on Olympiads, each year having a designation according to the formula 'the 2nd year of the

170th Olympiad' (= 99 BC). This system of dating was still in use in Byzantine times.

14.5   A system of consecutive counting from the year of the foundation of Rome (AUC, that is, *Ab Urbe Condita*) is attested in the time of Cicero (see *Brutus* 72, where the phrase is *post Romam conditam*); but scholars in his time were not agreed on the exact date of that foundation. Polybius had placed it in the second year of the seventh Olympiad (751 BC: compare Cicero: *Republic* II 18); Varro chose the third year of the sixth Olympiad (754 BC). This uncertainty was presumably part of the reason why this means of designating years was never widely used in Rome.

14.6   The modern method of designating years by consecutive counting AD or BC was instituted in AD 525.

B. MONTHS AND DAYS

14.7   The lunar month is the time taken by the moon to go through all its phases; its length varies somewhat but always lies between 29 and 30 days. It provides an obvious means of subdividing the year, but its use for this purpose brings some difficult problems in its train if the calendar is to fulfil its prime function in the classical world of ensuring that the religious festivals come in the right season each year. The main problem is that there is not an exact number of lunar months in the year: 12 lunar months = 354 days, leaving approximately 11 1/4 days to be filled in before the end of the year.

14.8   The Athenian Festival Calendar used lunar months of 29 and 30 days alternating, and attempted to maintain correlation with the seasons by inserting an extra month into the year every so often. But this intercalation was not consistently carried out, so that the months were often out of step with the seasons. Herodotus explains: 'Seventy years [of 30 days] are equivalent to 25200 days without an intercalary month (μὴν ἐμβόλιμος); and if it is desired that every second year (τὸ ἕτερον ἔτος) should be longer by a month so that the seasons should come round at the right time, there will be 35 intercalary months in the seventy years, a total of 1050 days' (I 32 3; but this is equivalent to a year of 375 days).

14.9   A more scientific attempt to deal with the problem was made by the astronomers Meton and Euctemon in 432 BC. They proposed an equation 19 years = 235 months and worked out a rather complicated calendar based on this 19-year cycle (still known as the Metonic cycle), which began on the day of the summer solstice in 432 BC (27 June by our reckoning). A slight error was corrected by Callippus in the fourth century BC by omitting the last day of every fourth cycle. This Callippic cycle was the one used by Hellenistic astronomers including Hipparchus, but it apparently never found its way into Athenian civil life. It is equivalent to a year of exactly 365 1/4 days.

14.10   The calendar of the Roman republic, ascribed by the Romans to King Numa, used twelve months which had presumably in origin been lunar, for a total of 355 days. As with the Athenian Festival Calendar (see 14.8), therefore, intercalation was necessary, but was not consistently carried out: a lunar eclipse which we know to have taken place on 21 June 168 BC is placed by Livy (XLIV 37) on the equivalent of 3 September. By 46 BC the calendar was 90 days out of step with the seasons; and at this stage Julius Caesar gave to the Alexandrian astronomer Sosigenes the task of reforming it.

14.11   Sosigenes started with the Egyptian calendar year of 365 days, which he brought closer to the solar year by adding one day to each fourth year (see 14.13). He abandoned lunar months in favour of twelve calendar months adding up to 365 days (or 366 in leap year). He retained the names of the months in the Republican calendar (but *Quintilis* was renamed *Iulius* in 44 BC and *Sextilis* became *Augustus* in 8 BC). He prolonged the year 46 BC (708 AUC) until the calendar came back into line with the seasons (that is, until the calendar year ended in midwinter); this year reached an eventual total of 445 days. The first year of the calendar now known as the Julian calendar was 45 BC.

14.12   The Athenians counted the days within their lunar months with an eye to the phases of the moon. The month was divided into three parts. The first part was the period of the waxing moon; the first day was νουμηνία 'new-month day' and then the counting started with δευτέρα ἱσταμένου 'the second (day) of the waxing (moon: μηνός)' and went up to δεκάτη ἱσταμένου 'the tenth of the waxing'. Over the next ten-day period the dependent genitive participle was dropped and the counting went up through ἑνδεκάτη 'eleventh', δωδεκάτη 'twelfth' and τρίτη ἐπὶ δέκα 'thirteenth' to δεκάτη ἐπὶ δέκα 'twentieth'. The remaining days of the month were the period of the waning moon, and the counting went backwards from δεκάτη φθίνοντος 'the tenth of the waning' (which was the 21st day of the month) to δευτέρα φθίνοντος 'the second of the waning' (the 29th of the month), and the last day of the month was called ἔνη καὶ νέα 'the old and new day'. In a 29-day month δευτέρα φθίνοντος was left out. It was on ἔνη καὶ νέα that interest was payable, as Strepsiades was uncomfortably aware (Aristophanes: *Clouds* 1134ff).

14.13   The Romans recognised three fixed points within the month and counted down to each of these in turn. They were the Kalends (1st of the month), Nones (5th of the month, but 7th in March, May, July and October), and Ides (eight days after the Nones, and hence 13th/15th of the month). In leap year 24 February was repeated; as, by the Roman reckoning, this day was the sixth (*sextus*) before the Kalends of March, the Latin for 'leap year' is *annus bisextilis* (and the Latin for 'leap year day' is *dies bisextus*). The full phrase expressing this date in Latin was *ante diem sextum Kalendas Martias* (see 3.3-4), abbreviated as *a. d. VI Kal. Mart.* or more simply *VI Kal. Mart.* The day before each fixed point was called *pridie*; hence 14 March = *pridie Idus Martias* or *prid. Id. Mart.* Table 14.1 (see pp. 64-5) sets out this calendar in detail.

C. HOURS

14.14   According to Herodotus (II 109 3) it was the Babylonian astronomers who invented the system of dividing day and night into twelve sections. His word for the sections is μέρη 'parts', not ὧραι 'hours'. In classical Greek ὧρα denoted any (recurrent) period of time; it meant much the same as the English word 'time' in such phrase as 'the time of day' or 'at this time of year'. It was the regular Greek word for the seasons of the year, and the year itself might be called a ὧρα. In the meaning 'hour' it is only post-classical. It was borrowed into Latin as *hora*; in Latin its regular meaning is 'hour'.

14.15   The division of day and night into twelve was maintained in all seasons, with the result that the daylight hours were longer in summer than in winter. Only at the equinoxes were the daylight and night-time hours of the same length, and then they coincided in length with the modern hour.

14.16   It appears to have been Hipparchus (second century BC) who first conceived the idea of applying the division into 24 equinoctial hours to every day in the year so that the hours were always the same length whatever the season. The idea was taken up only for scientific purposes: technical writers who wished to use equinoctial hours were always at pains to explain this to their readers, who would otherwise have assumed that the hours were the variable hours with which they were familiar. So Pliny writes (*Natural History* XVIII 221): *horae nunc in omni accessione ac decessione aequinoctiales, non cuiuscumque diei, significantur* 'in all addition and subtraction "hours" means "equinoctial hours", not those of any particular day'.

14.17   The daylight hours were numbered 1-12 starting at sunrise, and the night hours likewise starting from sunset. The ordinal numbers (see 2.1-2) were used in agreement with *hora*, and day or night was specified if necessary; so Sextus Roscius was murdered *post horam primam noctis* 'after the first hour of night (Cicero: *Pro Roscio Amerino* 19). The Latin for 'What is the time?' is *quota hora est?* 'Which hour is it?' and the answer took the form *tertia hora est* 'It is the third hour.' The phrase *tertia hora diei* 'the third hour of the day' had two meanings. It could refer to the whole period 8-9 a.m. (at the equinox); the hours were not further sub-divided in everyday usage, and any time between 8 and 9 a.m. would be referred to as 'the third hour of the day'. But it might also refer more specifically to the moment when that hour ended and the next began. Thus when Horace says *ad quartam iaceo* 'I lie (in bed) to the fourth hour' (*Satires* I 6 122) he may mean that he gets up at 10 a.m. (at the equinox).

14.18   The κλεψύδρα (in Latin *clepsydra*) was a water-clock similar in function to our hourglass: it was used, not for telling the time, but for measuring fixed periods of time, and its main use was to regulate the length of speeches in the law courts. The measurement of time by a *clepsydra* was expressed not in hours and minutes but in terms of the volume of water that passed through the clock. In Demosthenes the phrase ἐν τῷ ἐμῷ ὕδατι 'in my water' means 'in the amount of time allotted to me' (XIX 57). Elsewhere he says 'the archon had to pour in [that is, into the *clepsydra*] 40 litres for each of the parties and 10 litres for the final speech' (XLIII 8). The younger Pliny writes: 'I spoke for nearly five hours; for I had been allotted twelve of the largest *clepsydrae*, and four had been added to these' (so the *clepsydra* was filled sixteen times: *Letters* II 11 14).

D. MINUTES AND SECONDS

14.19   At some period after 450 BC the Babylonians divided the zodiac into 360 degrees; and using their system of sexagesimal division they also divided each degree into 60. The Hellenistic astronomers referred to these sixty subdivisions as λεπτά 'small parts', and this word was later translated into Latin as *minuta*, giving us our word 'minutes'. A further division of each λεπτόν into 60 produced δεύτερα λεπτά 'second (set of) small parts'; these became in Latin *secunda minuta*, which gave us our word 'seconds'.

14.20   These subdivisions of the hour were used only by scientists in the classical world and were never part of everyday life. The elder Pliny (*Natural History* II 58) divides the hour by means of the *uncia* fractions (see 5.9), expressing as *dodrans semuncia horae* what we would express as 47 1/2 minutes.

	Jan.	Feb.	March	April	May	June
1	Kal.	Kal.	Kal.	Kal.	Kal.	Kal.
2	IV	IV	VI	IV	VI	IV
3	III	III	V	III	V	III
4	prid.	prid.	IV	prid.	IV	prid.
5	Non.	Non.	III	Non.	III	Non.
6	VIII	VIII	prid.	VIII	prid.	VIII
7	VII	VII	Non.	VII	Non.	VII
8	VI	VI	VIII	VI	VIII	VI
9	V	V	VII	V	VII	V
10	IV	IV	VI	IV	VI	IV
11	III	III	V	III	V	III
12	prid.	prid.	IV	prid.	IV	prid.
13	Id.	Id.	III	Id.	III	Id.
14	XIX	XVI	prid.	XVIII	prid.	XVIII
15	XVIII	XV	Id.	XVII	Id.	XVII
16	XVII	XIV	XVII	XVI	XVII	XVI
17	XVI	XIII	XVI	XV	XVI	XV
18	XV	XII	XV	XIV	XV	XIV
19	XIV	XI	XIV	XIII	XIV	XIII
20	XIII	X	XIII	XII	XIII	XII
21	XII	IX	XII	XI	XII	XI
22	XI	VIII	XI	X	XI	X
23	X	VII	X	IX	X	IX
24	IX	VI	IX	VIII	IX	VIII
25	VIII	V	VIII	VII	VIII	VII
26	VII	IV	VII	VI	VII	VI
27	VI	III	VI	V	VI	V
28	V	prid.	V	IV	V	IV
29	IV		IV	III	IV	III
30	III		III	prid.	III	prid.
31	prid.		prid.		prid.	

Table 14.1: The Roman calendar year

	July	Aug.	Sept.	Oct.	Nov.	Dec.
1	Kal.	Kal.	Kal.	Kal.	Kal.	Kal.
2	VI	IV	IV	VI	IV	IV
3	V	III	III	V	III	III
4	IV	prid.	prid.	IV	prid.	prid.
5	III	Non.	Non.	III	Non.	Non.
6	prid.	VIII	VIII	prid.	VIII	VIII
7	Non.	VII	VII	Non.	VII	VII
8	VIII	VI	VI	VIII	VI	VI
9	VII	V	V	VII	V	V
10	VI	IV	IV	VI	IV	IV
11	V	III	III	V	III	III
12	IV	prid.	prid.	IV	prid.	prid.
13	III	Id.	Id.	III	Id.	Id.
14	prid.	XIX	XVIII	prid.	XVIII	XIX
15	Id.	XVIII	XVII	Id.	XVII	XVIII
16	XVII	XVII	XVI	XVII	XVI	XVII
17	XVI	XVI	XV	XVI	XV	XVI
18	XV	XV	XIV	XV	XIV	XV
19	XIV	XIV	XIII	XIV	XIII	XIV
20	XIII	XIII	XII	XIII	XII	XIII
21	XII	XII	XI	XII	XI	XII
22	XI	XI	X	XI	X	XI
23	X	X	IX	X	IX	X
24	IX	IX	VIII	IX	VIII	IX
25	VIII	VIII	VII	VIII	VII	VIII
26	VII	VII	VI	VII	VI	VII
27	VI	VI	V	VI	V	VI
28	V	V	IV	V	IV	V
29	IV	IV	III	IV	III	IV
30	III	III	prid.	III	prid.	III
31	prid.	prid.		prid.		prid.

Table 14.1: The Roman calendar year

SELECT BIBLIOGRAPHY

The books in this list are referred to in the text by means of the author's surname

Bickerman, E.J., *Chronology of the Ancient World* (London, Thames and Hudson, revised edition 1980). Covers in much greater detail the material of chapter fourteen.

Bulmer-Thomas, I., *Greek Mathematical Works* (Loeb Classical Library, London, Heinemann, Volume I 1939, Volume II 1941, repr. 1991-3). Greek texts with English translation on facing page. Volume I has material about square roots and Volume II includes Archimedes' *Arenarius* and *Measurement of the Circle*.

Crook. J.A., *Law and Life of Rome* (London, Thames and Hudson, 1967). Includes discussion of legal matters pertaining to loans.

Dilke, O.A.W., *The Roman Land Surveyors* (Newton Abbot, David and Charles, 1971). Relevant to chapters seven and eight.

Head, B.V., *Historia Numorum: a Manual of Greek Numismatics* (London, Spink, 1963, reprint). Standard work on Greek coinage.

Hodge, A. Trevor, *Roman Aqueducts and Water Supply* (London, Duckworth, 1992). Relevant to chapter thirteen.

Hultsch, F., *Scriptores Metrologici* (Stuttgart, Teubner, reprint 1971). Collection of ancient testimonia on weights, measures and coinage, with nearly two hundred pages of Prolegomena in Latin. Brings together much material otherwise difficult of access. Relevant to chapters seven to ten.

Landels, J.G., *Engineering in the Ancient World*, (London, Chatto and Windus, 1978). His second chapter provides background material for chapter thirteen.

MacDowell, D.M., *The Law in Classical Athens* (London, Thames and Hudson, 1978). Includes discussion of legal matters pertaining to loans.

Tod, M.N., *Ancient Greek Numerical Systems* (Chicago, Ares,1979). Examines in great detail the inscriptional evidence for Greek numerals.

Woodhead, A.G., *The Study of Greek* Inscriptions (Cambridge University Press, 1981). Relevant to chapter four.

INDEX OF PASSAGES QUOTED

GREEK AUTHORS

		II 175	7.3
		II 180 1	5.2b
		III 95 2	6.1
		V 88 2	5.13
		V 89 2	2.2
		VI 63 1	3.2
		VI 63 2	6.6
		VII 48	6.3a
		VII 186 2	1.4
Hippocrates:	*Muliebria*	34	5.16
Homer:	*Iliad*	VI 236	11.1
		XXIII 751	5.1a
Plato:	*Apology*	36b 1	5.2a
	Laws	742c	12.2
		895e 3	6.4a
	Meno	82c 5	8.2
	Theaetetus	147e-148a	8.2
		148b 2	8.8
		175b 1	2.2
	Timaeus	35b 4	6.3a
Plutarch:	*Gaius Gracchus*	7	7.9
	Lycurgus	12	5.16
Pollux:	*Onomastikon*	II 158	7.3
Polybius:	*Histories*	V 26 13	6.7
		XXI 43 19	9.5
Strabo:	*Geography*	VII 7 4	7.9
Thucydides:	*History*	I 10 2	5.4a
		I 108 2	2.2
		II 13 3	2.11
		II 49 6	2.6
		II 78 2	5.1b
		II 97 1	2.6
		IV 38 5	2.11
		IV 83 5	5.1c
		IV 102 3	2.11

	De Oratore	II 109	6.2
	De Republica	II 18	14.5
	In Verrem	I 92	1.12
		II 122	2.5
		III 165	12.5
	Orator	188	5.14
	Philippics	II 93	1.13
	Pro Cluentio	87	1.11
		179	9.3
	Pro Flacco	30	1.12
	Pro Plancio	62	5.8
	Pro Roscio Amerino	19	14.15
Columella:	*De Re Rustica*	V 1 6	7.10; 8.3
		V 2	8.7
		V 2 1	6.3d
		V 2 2	8.5
		V 2 5	8.3
Frontinus:	*Aqueducts*	I 25	13.2
		I 39-63	13.6f
		I 62-3	13.5
Horace:	*Ars Poetica*	325-30	5.9
	Odes	III 21 1	14.3
	Satires	I 2 14	12.5
		I 6 122	14.15
		II 5 100-1	12.10
Juvenal:	*Satires*	III 61	5.8
		XIII 95	5.5d
Leges Publicae:			10.11
Livy:	*History of Rome*	VI 15 10	12.2
		VII 8 1	6.3c
		VII 16 1	2.10
		VII 19 5	12.5
		VII 27 3	2.10
		X 47 2	2.2
		XXI 47 8	7.5

Tacitus:	*Annals*	VI 16	2.10; 12.5
	Histories	II 43	2.7
Varro:	*De Lingua Latina*	V 173	2.8
		V 174	11.8
	De Re Rustica	I 10 2	8.6
Virgil:	*Aeneid*	X 565-6	2.5
Vitruvius:	*Architecture*	VIII 6 4	13.2
		IX Pref. 9-12	8.8
		IX Pref. 4	8.3
		IX Pref.13	8.8

INDEX OF WORDS

References are to paragraphs unless otherwise indicated

ENGLISH

LATIN

congius, 10.5-6
cyathus, 10.6

decempeda, 7.6
denarius, 2.8; 9.7 ff; 11.7 ff
digitus, 7.1; 7.4; 13.4
dimidium, 5.5c
dimidius, 5.5d
duodeuiginti, 2.11
dupondius, 11.12

faeneror, 12.2
faenus, 12.2; 12.5
fistula, 13.1 f

hemina, 10.5-8
heredium, 8.6
hora, 14.14

iugerum, 8.4; 8.6

libella 5.12; 11.8; 12.11
libra, 9.3; 10.8

medimnus, 10.10
mille, 1.10; 4.5-6
minutum, 14.19
modius, 10.10
modulus, 13.1
multiplico, 6.3d

pars, 5.6-7
passus, 7.5
pertica, 7.6
pes, 7.1

Printed and bound by CPI Group (UK) Ltd, Croydon, CR0 4YY

27/10/2024

14580407-0002